Rourke
Educational Media
rourkeeducationalmedia.com

Animals Have Classes Too!

Fish

Christa C. Hogan

Before, During, and After Reading Activities

Before Reading: Building Background Knowledge and Academic Vocabulary

"Before Reading" strategies activate prior knowledge and set a purpose for reading. Before reading a book, it is important to tap into what your child or students already know about the topic. This will help them develop their vocabulary and increase their reading comprehension.

Questions and activities to build background knowledge:
1. Look at the cover of the book. What will this book be about?
2. What do you already know about the topic?
3. Let's study the Table of Contents. What will you learn about in the book's chapters?
4. What would you like to learn about this topic? Do you think you might learn about it from this book? Why or why not?

Building Academic Vocabulary

Building academic vocabulary is critical to understanding subject content.
Assist your child or students to gain meaning of the following vocabulary words.

Content Area Vocabulary
Read the list. What do these words mean?

- cartilage
- evolved
- parasites
- phylum
- resemble
- spawn
- species
- venomous

During Reading: Writing Component

"During Reading" strategies help to make connections, monitor understanding, generate questions, and stay focused.
1. While reading, write in your reading journal any questions you have or anything you do not understand.
2. After completing each chapter, write a summary of the chapter in your reading journal.
3. While reading, make connections with the text and write them in your reading journal.
   a) Text to Self – What does this remind me of in my life? What were my feelings when I read this?
   b) Text to Text – What does this remind me of in another book I've read? How is this different from other books I've read?
   c) Text to World – What does this remind me of in the real world? Have I heard about this before? (News, current events, school, etc.…)

After Reading: Comprehension and Extension Activity

"After Reading" strategies provide an opportunity to summarize, question, reflect, discuss, and respond to text. After reading the book, work on the following questions with your child or students to check their level of reading comprehension and content mastery.
1. What are the three classes of fish? *(Summarize)*
2. What kinds of advantages might cartilage gives sharks and rays over bony fish? *(Infer)*
3. How are a lamprey and a salmon alike? How are they different? *(Asking Questions)*
4. If you had to make a new way to classify fish, what rules would you make? *(Text to Self Connection)*

Extension Activity
Cut pictures of fish from magazines. Sort them into groups. What do the fish in each group have in common? What makes them different?

Table of Contents

Let's Classify! 4
Fish Classes 8
Bony Fish. 10
Jawless Fish 12
Sharks, Rays, and Skates 16
Activity 21
Glossary 22
Index 23
Show What You Know 23
Further Reading. 23
About the Author 24

stingray

Let's Classify!

Earth is home to many **species** of living things. Scientists group them together by common traits. This is called classification. Classification helps scientists understand how species are related.

Animal Groups

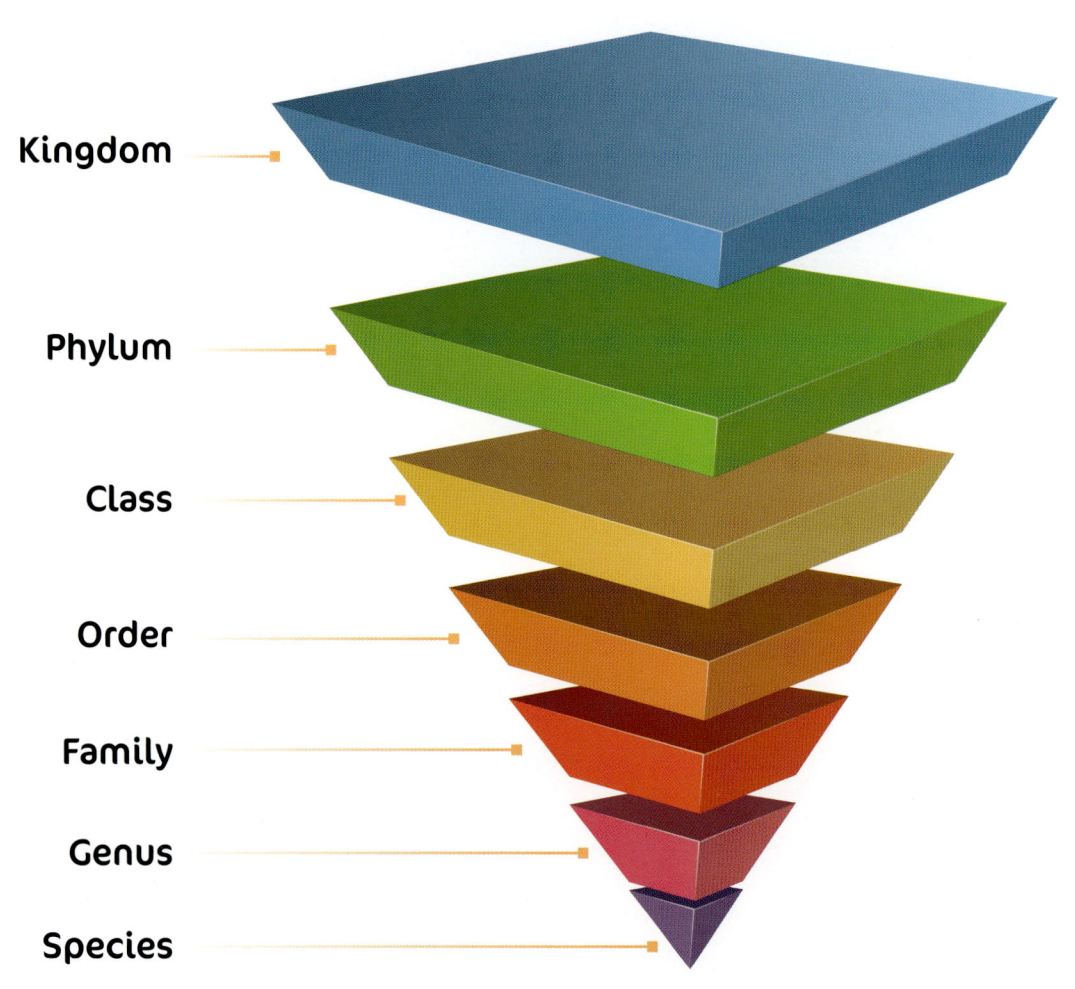

- Kingdom
- Phylum
- Class
- Order
- Family
- Genus
- Species

All species are divided into six kingdoms. Each kingdom is separated into smaller groups.

Aquarium visitors can view a variety of saltwater fish, including sharks and rays.

Scientists break the animal kingdom into two phyla: vertebrates and invertebrates. Vertebrates have backbones. Invertebrates do not. Fish are in the vertebrates **phylum**.

Phyla are divided into smaller groups called classes. There are three classes of fish. They are: bony fish; jawless fish; and sharks, rays, and skates. Each class is divided into orders, families, genera, and species.

A skate's mouth is well suited for burrowing in the sand in search of food.

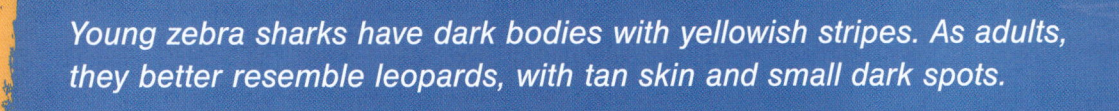

Young zebra sharks have dark bodies with yellowish stripes. As adults, they better resemble leopards, with tan skin and small dark spots.

The First Classification

Carl Linnaeus first used classifications in 1753. Scientists have now classified 1.2 million species. They think there are still more than 8 million species yet to be discovered!

Fish Classes

Scientists place animals in the fish class when they share a few common traits. Fish have scales. They use gills to breath underwater. They have fins that help them move. All fish are cold-blooded. Their bodies are the same temperature as the water.

scales — *dorsal fin* — *caudal fin* — *gill* — *pelvic fin* — *anal fin*

Fish pass water through their gills to remove oxygen. This helps them breathe. Fins help fish move through the water.

Fish Habitats
Fish live in salt water, fresh water, or a mix of both. Water that is both salt and fresh water is called brackish. Brackish water is found where fresh water, such as a river, meets the ocean.

The fish in each smaller class also share common traits. These groups help scientists organize new species.

Scientists have discovered more than 29,000 species of fish.

Bony Fish

Bony fish have hard bones. Their skin is made of many overlapping scales. Bony fish live in fresh or saltwater habitats. Bony fish also **spawn** eggs in the water.

One salmon lays between one thousand and ten thousand eggs.

Lots of Fish!
More than 26,000 species belong to the bony fish class. Bony fish species account for more than half of all known vertebrates!

Salmon are bony fish. They are born in freshwater streams and live their adult lives in the ocean. They return to their birthplace to lay eggs.

Humphead wrasses are also bony fish. These giants are 7 feet (2 meters) long and weigh 420 pounds (190 kilograms). If the male wrass leaves the school, the largest female wrass becomes male.

Humphead wrasses are known as "elephants of the coral reef." Fish that change sex like wrasses are called hermaphrodites.

Jawless Fish

Jawless fish are living dinosaurs. Scientists discovered jawless fish fossils from 500 million years ago. Jawless fish were plentiful then. Now only lampreys and hagfish remain.

Jawless fish were the first fish to appear in fossil records.

Jawless fish **resemble** eels. They have long, thin bodies without fins, scales, or jaws. They have sucker-like mouths with circles of teeth.

This sea lamprey uses its teeth to feed off of other fish, such as sharks.

Hunting by Scent
Some jawless fish, like hagfish, lack strong eyes. They use their sense of smell to hunt.

Lampreys haven't changed much in more than 340 million years. Most lampreys are **parasites**. They latch onto other fish and eat their blood and tissue. The host fish often dies from its wounds.

Sea lampreys have invaded several of the Great Lakes. One sea lamprey kills up to 40 pounds (18 kilograms) of fish in one season.

Hagfish use toothed tongues to eat dead and injured fish. Hagfish are also known as slime fish.

Hagfish produce slime that helps them escape from predators during an attack.

Slimy Suckers
Slime helps the hagfish burrow inside injured fish. The hagfish then eats the fish from the inside out.

Sharks, Rays, and Skates

Sharks **evolved** 420 million years ago. There are now more than 440 known species of sharks. Sharks are fish, but they have rough skin instead of scales. Their bones are made of flexible, light **cartilage**.

Rays and skates have wide, flat bodies with wings. To swim, they beat their wings in a flying motion like birds. They also have long, thin tails. Some rays have venomous barbs on their tails, while skates do not.

Great white sharks can grow up to 20 feet (6 meters) and weigh 2.5 tons (1,800 kilograms).

mermaid's purse

Real Mermaids
Rays and skates look alike, but rays produce live young. Skates lay eggs or "mermaid's purses."

Sharks, rays, and skates use electroreception to hunt. They sense the bioelectric fields in the nervous systems of their prey.

Whale sharks can live for 130 years and grow 60 feet (18 meters) long. That's three times bigger than a great white shark. However, whale sharks are gentle giants. They feed on plankton and small fish.

Whale sharks swim with their mouths open to collect plankton, shrimp, and krill.

The shy blue-spotted ribbontail ray will only use its barbed tail to defend itself when threatened.

Blue-spotted ribbontail rays are well-defended from predators. Their barbed tails have a **venomous** sting. They also avoid predators by hiding along the sandy ocean floor.

Scientists discover thousands of new species each year. Since underwater habitats have been difficult to study in the past, many of these new species are fish.

In 2017, researchers discovered a large freshwater stingray. The *Potamotrygon rex* is 43 inches (109 centimeters) long. It weighs 44 pounds (20 kilograms). Scientists also found a new snail fish living in the deepest part of the world's oceans.

A scientist uses an aqua scope to observe life along the edges of the Baltic Sea.

Scientists continue to learn more about fish. Classification helps them understand where new fish fit in the animal kingdom.

ACTIVITY

Fish Food

Fish have many ways to hunt for food. Some fish use echolocation. They use sound to locate food. Test your echolocation skills with a fishy version of "Marco Polo."

Supplies

blindfold

two or more players

Directions

1. One blindfolded person acts as the "Fish." All other players are "Food."

2. When the Fish says, "Fish," all players reply, "Food." The fish must use only his hearing to capture Foods.

3. The first Food to be captured becomes the next Fish.

Glossary

cartilage (KAHR-tuh-lij): a strong, elastic tissue that forms the outer ear and nose of humans and mammals, and lines the bones at the joints

evolved (i-VAHLV): to change slowly and naturally over time

parasites (PAR-uh-site): animals or plants that live on or inside of another animal or plant

phylum (FYE-luhm): a group of related plants or animals that is larger than a class but smaller than a kingdom

resemble (ri-ZEM-buhl): to look like or be similar to something or someone

spawn (spawn): to produce a large number of eggs

species (SPEE-seez): a group of living things of the same kind with the same name

venomous (VEN-uhm-us): containing poison, as in the bites of some snakes and spiders

sawfish

Index

bony fish 6, 10, 11
hagfish 12, 13, 15
jawless fish 12, 13
Linnaeus, Carl 7
salmon 10, 11

shark(s) 5, 6, 7, 16, 17, 18
skates 6, 16, 17
vertebrates 6

Show What You Know

1. How does classification help scientists?
2. What is the difference between a shark, a ray, and a skate? What do they have in common?
3. Name three physical traits that help scientists identify fish.
4. Which fish are known as "living dinosaurs"?
5. How many species of fish have been discovered?

Further Reading

Fretland VanVoorst, Jennifer, *Animal Classification*, Abdo, 2014.

Royston, Angela, *Fish*, Heinemann, 2015.

Sanchez, Anita, and Stock, Catherine, *Karl, Get Out of the Garden!: Carolus Linnaeus and the Naming of Everything*, Charlesbridge, 2017.

About the Author

Christa C. Hogan is an author and a PADI-certified scuba diver. She grew up watching Jaques Cousteau on television. Her favorite dive experience was spotting a small octopus in Costa Rica. She dreams of one day diving with a whale shark. For now, she's happy to be writing about them. She also loves narwhals and cephalopods.

Meet The Author!
www.meetREMauthors.com

© 2019 Rourke Educational Media

All rights reserved. No part of this book may be reproduced or utilized in any form or by any means, electronic or mechanical including photocopying, recording, or by any information storage and retrieval system without permission in writing from the publisher.

www.rourkeeducationalmedia.com

PHOTO CREDITS: Cover and Title Pg ©LeventKonuk; Border ©cinoby; Pg 3 ©aon168; Pg 4 ©lvcandy; Pg 5 ©themacx; Pg 6 ©JovanaMilanko; ©jeffhochstrasser; Pg 7 ©Placebo365; Pg 8 ©7activestudio; Pg 9 ©Rainer von Brandis; Pg 10 ©arctic-tern; Pg 11 ©flyingrussian;Pg 12 ©imagoRB; Pg 13 ©PEDRE; Pg 14 ©Arsty; Pg 15 ©ffennema; Pg 16 ©LeicaFoto; Pg 17 ©Nigel_Wallace; Pg 18 ©WhitcombeRD; Pg 19 ©neosummer; Pg 20 ©MaslennikovUppsala; Pg 21 ©eliflamra; Pg 22 ©inguaribile

Edited by: Keli Sipperley
Cover and interior design by: Kathy Walsh

Library of Congress PCN Data

Fish / Christa C. Hogan
(Animals Have Classes Too!)
ISBN 978-1-64369-031-5 (hard cover)
ISBN 978-1-64369-107-7 (soft cover)
ISBN 978-1-64369-178-7 (e-Book)
Library of Congress Control Number: 2018956077

Rourke Educational Media
Printed in the United States of America,
North Mankato, Minnesota